Earth Changes

Dona Herweck Rice

Publishing Credits

Rachelle Cracchiolo, M.S.Ed., *Publisher*
Conni Medina, M.A.Ed., *Managing Editor*
Nika Fabienke, Ed.D., *Content Director*
Véronique Bos, *Creative Director*
Shaun N. Bernadou, *Art Director*
Carol Huey-Gatewood, M.A.Ed., *Editor*
Valerie Morales, *Associate Editor*
Courtney Roberson, *Senior Graphic Designer*

Image Credits: All images from iStock and/or Shutterstock.

Teacher Created Materials
5301 Oceanus Drive
Huntington Beach, CA 92649-1030
www.tcmpub.com
ISBN 978-1-4938-9818-3
© 2019 Teacher Created Materials, Inc.

It was ☀.

day

It is 🌙.
night

It was 🌡.
cold

It is 🌡.
warm

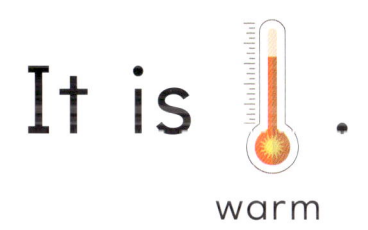

It was 🪨.

dry

It is 💧.
wet

It was 🟢.
green

It is ●.
brown

It was 🌱.
young

It is 🌳.
old

High-Frequency Words

is

it

was